The Science of
Angels

Peter Walling

authorHOUSE®

AuthorHouse™
1663 Liberty Drive
Bloomington, IN 47403
www.authorhouse.com
Phone: 1 (800) 839-8640

Published by AuthorHouse   02/06/2017

ISBN: 978-1-5246-7034-4 (sc)
ISBN: 978-1-5246-7032-0 (hc)
ISBN: 978-1-5246-7033-7 (e)

Library of Congress Control Number: 2017901715

Print information available on the last page.

Contents

Contents

Dedication

This volume is dedicated to those bridge builders who would attempt to span the chasm between science and religion.

Preface

The title of this book is intended to be provocative, but not anarchic. Modern science and ancient philosophy need not be mutually antagonistic; they are both products of the skill and ingenuity of Homo sapiens and both deserve the respect which ought to be accorded to the contributors. The author is trained in physiology, medicine and anesthesiology; theological knowledge was gleaned from an Anglican boarding school in England and attendance at The

Episcopal Church of The Annunciation in Lewisville, Texas. Any theological blunders are the responsibility of the author and not his masters or pastors.

The appreciation of Angels is a sadly neglected part of Church teaching and worship. This state of affairs is probably due to their apparently impossible appearance and subsequent disappearance which must strike the modern thinker as being like science fiction. In addition to this, the appearance of Angels is rare and unpredictable so that investigators suffer problems associated with any rare event be it an asteroid impact or the parting of the Red Sea.

The main job description of an Angel is to act as a messenger of God; this in its turn requires a sentient creature to receive the message. This brings us to the next problem; not only are angelic operations mysterious, but the nature of human consciousness remains unknown.

During the past 30 years great strides have been made in understanding the nature of consciousness (1, 2). One of the main advances has been the better understanding of non-linear dynamics and chaos theory. At Baylor University Medical Center in Dallas we studied patients as they emerged from anesthesia and described the non-linear response to diminishing levels of anesthetic, and for the first time demonstrated stepwise changes in neurodynamics as consciousness returned (3). These studies paved the way to a better understanding of the way in which the dynamics of the human brain utilize fractions of dimensions and higher dimensions up to about 5D. It occurred to us that the same arguments used to explain consciousness might be extended to help explain angelic appearances. This short volume is a companion to "Consciousness: Anatomy of the Soul" (4). We make no apologies for copying some of the figures as this text would be hard to understand otherwise. For a better explanation of fractals, attractors and dimensions, the reader is encouraged to refer back to the older work.We deal with consciousness, angels and finally how the two may connect.

We have used the 1611 King James Version (KJV) of the bible because of the beauty of the text and the unchallenged integrity and scholarship of the translators. Descriptions of angels are taken at face value; we hold the writings of our distant ancestors in very high regard. Additionally we make no apologies for quoting writers from the pre scientific era. The ~750 year old Summa Theologicae of St.

Thomas Aquinas was written at a time when scholars were taught to think and argue, and ideas about angels were consisered to be extremely important(5).

CHAPTER 1

Consciousness

A mother takes her six year old child to the zoo. Tatum has never been to the zoo before but at age 6 she knows what most of the animals look like from books and TV. Tatum is 3.5 feet tall and weighs 50 pounds. Mother and child stop opposite the elephant enclosure. An elephant stands 20 feet away separated by a low wall and a moat. The elephant is 12 feet tall and weighs 5 tons. Tatum looks at the elephant.

Let us consider what happens next. Like it or not, we all live in a swirling sea of electromagnetic radiation. These include radio waves, X-rays, microwaves, cell phone transmissions and light waves. Some pass through us and others bounce off. It would be a waste to process them all but as humans, we have evolved the capacity to process wavelengths from about 400-700 nanometers (billionths of a meter). This is the visible light

3

spectrum from violet to red. Light from the elephant is focused by the lens in our eye onto the inside of the back of the eyeball, the retina. The image is upside down but that is not the strangest thing. The cells of the photoreceptors covert the electromagnetic radiation to nerve signals. Elephant data is digitized into a staccato stream of impulses racing along the million fibres in each optic nerve, in total darkness. Note well that the elephant did not enter Tatum's brain, coded impulses did. The nerve impulses travel backwards via the optic chiasm to form the optic tract around the midbrain and on to the lateral geniculate nucleus (LGN), then to the optic radiation and on to the primary visual cortex. The visual signals are processed here and the next thing Tatum knows is that the elephant appears in her visual perceptual space. Tatum is the percipient, but what and where is the percept, i.e. the elephant?

Milliseconds before the percept appeared all the information describing the elephant consisted of nerve signals; the brain did not construct a flesh and blood pachyderm! Tatum's neurons exist in physical space but the percept itself is not a physical object in physical space; it is a non-physical construct in non-physical space. The percept is also scale free, it has no size. A 5 ton elephant will not fit into the head of a 50 pound child.

The real, physical elephant is 20 feet away from Tatum and the visual image enjoyed by Tatum seems to correspond with that estimated distance. However, we know that the neural machinery producing the percerpt is in the brain, so is the percept inside or outside Tatum's brain?

The answer is neither.

To assign a location in physical space, either inside or outside the skull, to a non-physical object in non-physical space, is a non sequitur. As Bertrand Russell pointed out 90 years ago, "Physical and perceptual space have relations, but they are not identical, and failure to grasp the difference between them is a potent source of confusion" (6). An enormous amount of time and effort have been wasted in the search for the fabled anatomical correlate of consciousness.

We now need to consider some aspects of vertebrate brain evolution before we can understand what is going on. During the past 500,000,000 years we have evolved from fish like creatures into the Homo sapiens of today. The story is punctuated by giant leaps forward, the appearance of limbs, migration to dry land and mammalian thermoregulation being a few examples.

Generally speaking the brain has enlarged while retaining its position at the "sharp" end of the animal. Brain waves travel quite quickly but they are still slower than light speed by a factor of about 1,000,000. It therefore behoves the creature to keep the neurons coordinating the senses as compact as possible thus reducing communication speed. Whether it be predator or prey, an animal will benefit from the rapid sorting of incoming sensory signals.

During our animal studies we discovered an additional feature of the evolving vertebrate brain. Not only did the attractor dimension increase during evolutionary time but in humans, the dimension recorded correlated roughly with the number of senses being employed. (The attractor describes the moment to moment balance of forces involved in the neurodynamics and its measurement is reported elsewhere. (4) (Fig. 1). The attractor of a pendulum is a circle; as other forces enter the equation, the attractor becomes more complex and its dimension increases.

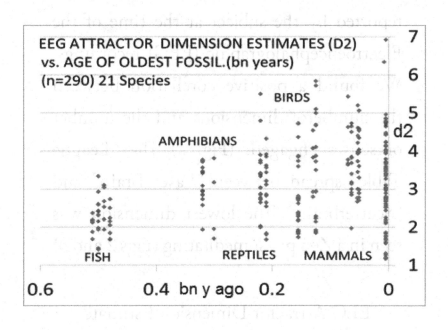

Figure 1. Attractor Dimensional Estimates increase during evolutionary time. Modern man appears on the right hand side with dimensions between about 1D and 5D. (Excluding outliers).Authors research.

In humans, we checked back in the records to discover the brain activity reported by the subject at the time of the Electroencephalographic (EEG) recordings. We found a positive corellation between the numberof dimensions and the number of senses engaged. (Fig. 2) The "League Table" spread between "Laser Brain" and "Scatterbrain". The lowest dimension was seen in a Zen priest meditating (Figs 2 and 3)

EEG Attractor Dimension Estimate

"Scatterbrain"

Multitasking. 4.8

Big Gestalt. 3.2

Small Gestalt. 2.8

Mental Math. 2.2

Burst of Volition. 2.1

Yoga Meditation. 2.0

Lords Prayer. 1.97

Zen Meditation 1.3

"Laser Brain"

Figure. 2 " League Table" shows increasing attractor dimension as more senses are incorporated. A gestalt is a combined sensory experience, for example seeing and hearing at the same time.

Notice that with mental arithmatic the brain is working hard but the mind is focussed hence the low dimension of 2.2. In the small gestalt, only two senses are involved and in the big gestalt, four senses are involved, hearing, vision, taste and smell.

Zen Priest . ~7 Hz.
~1.3 D Limit Cycle Attractor .

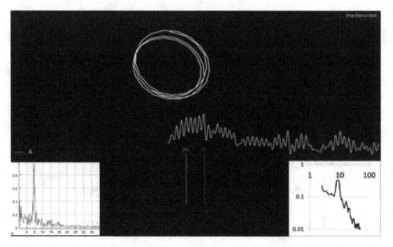

Akira KASAMATSMU.D, . and Tomio HIRAIM,D. Folia Psychiatrica et Neurologica. Japonica, Vol. 20, No. 4, 1966

Figure. 3 Electroencephalogram with associated attractor of Zen Priest meditating. A most unusual trace demonstraring an almost pure 7Hz signal. . (*Akira KASAMATSMU.D, . and Tomio HIRAIM, .D. Folia Psychiatrica et Neurologica. Japonica, Vol. 20, No. 4, 1966*)

Why has the brain evolved this way? Why does the brain need to employ non-physical space at all? The sensory system of an animal must gather the maximum information for survival at the least possible cost. We have considered Tatum's vision but what about the other senses? For example, vision senses electromagnetic radiation, the sense of smell involves chemical analysis while hearing analyses vibrations in the air or water surrounding the creature. Animals do not sample their environments sequentially; they would be eaten before they were half way through deciding where the predator was. They employ "binding", that is the bundling of information so that multiple senses may be processed at the same time. Our league table suggests that binding is accomplished by allocating approximately one dimemsion in the brain's neurodynamics to each sense. In this way visual information and auditory

13

information for example, may be appreciated simultaneously without one contaminating the other. The non-physical perceptual space provides common ground for this to happen. Non-physical space is also "elastic" enough to accommodate up to 5D. This would be impossible in physical space. The accumulation of sensory data in non-physical perceptual space helps to explain another conumdrum; "How is it possible for binding of information to occur when the brain areas processing the information are in different unconnected locations?" (Figs. 4, 5, 6).

Two possible carrier frequencies; 0.15-1.6 Hz, and 18-19 Hz. How may data be perceived simultaneously and yet retain separate qualities? (4/28/13/e 28-38 sec)

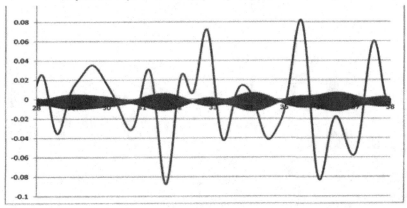

Figure 4. Two superimposed carrier waves from an EEG. For example hearing and vision. Author. Unpublished data.

Same signal with frequencies "fused". Mutual distortion?

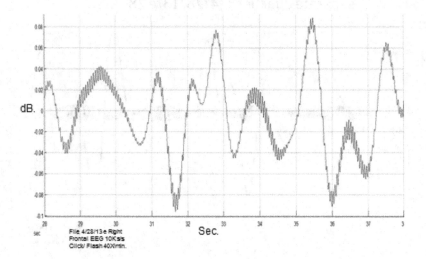

Figure 5. The same brainwaves as Fig. 4, fused and leading to mutual distortion. Author. Unpublished data.

The Natural State
"Separated" by 3D co-ordinate embedding.
Signals inhabit their own dimension.

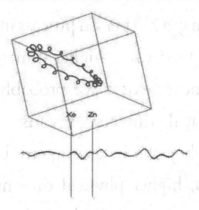

File 4/28/13 e Right
Frontal EEG 10Ks/s
Click/ Flash 400/min.

Figure 6. The two signals are separated by 3D co-ordinate embedding and presented in a more natural state. The slow component forms a lazy loop while the high frequency is represented by a tight coil at ~90 degrees. (i.e. in a higher dimension). In this way, although the two signals are "bound" together, they retain their individual properties. Author. Unpublished data.

The realization that perceptual space may exist in a higher dimensional immediately presents the next problem.

How may a 3D person process information that might exist in a higher dimension than they themeves exist? We probably live in a 3D physical universe because receeding galaxies obey the inverse square law. Small, curled up, higher physical dimensions may exist. However, in non-physical space higher dimensions are the norm but it's hard for us to visualize them. A tesseract is a four dimensional object where three squares meet at every edge; all we can perceive is its shadow.

To better understand higher dimensions we reduce them down to a more manageable level. We have adapted the story of Flatland (7). In this story all our usual dimensions are reduced by 1D so that we may better appreciate higher dimensional beings.

Dimensional Analogy.
How (*n*-1) dimensions relates to *n* dimensions.
Mr. Square in Flatland.

Walling PT, Hicks KN.
Consciousness:
Anatomy of the Soul. (2009)
From "Flatland: A Romance Of
Many Dimensions"
Edwin Abbott Abbott. 1884.

Figure 7. In Flatland, Mr. Square lives on a completely flat surface. (Authors' diagram.)

A flat square inhabits Flatland. Mr. Square cannot conceive of up or down. He lives in a flat house shaped like a polygon one side of which hinges to let him in and out.

Sphere visits Flatland

Figure 8. Once every 1,000 years a Sphere comes to visit. Mr Square cannot see the Sphere which is outside his event horizon.(Authors' diagram.)

Every 1,000 years, Flatland is visited by a 3D sphere whose mission is to convince Mr. Square that higher dimensions exist. As Sphere hovers over Flatland, he remains invisible to Mr. Square as he exists outside Mr. Square's event horizon. As soon as Sphere touches Flatland, Mr. Square is able to see Sphere, first as a dot, then as a disc when the 3D shape passes through the fabric of Flatland. The millennial visits are intended to allow the Sphere to persuade Mr. Square that higher dimensions exist, but here we hijack the story for our own ends.

Although Mr. Square is dimensionally deprived, his cunning knows no bounds. He knows that if the Sphere passes *through* Flatland an expanding hole will be created in the fabric of the surface which will be visible to him. Not only that, but if Mr. Square can remember the changing shapes and stack them up in his own perceptual space, he will form an image of the Sphere.

Sphere passes through the fabric of Flatland. Square sees a 2D disc.

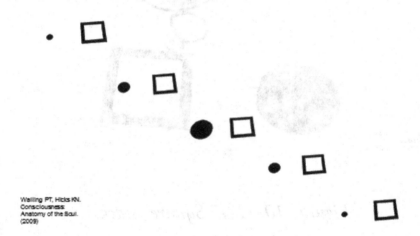

Figure 9. Sphere makes a hole in the fabric of Flatland which gets bigger then smaller as he passes through. (Authors' diagram).

2D Square constructs 3D image by stacking discs in perceptual space.

Figure 10. Mr Square stacks up the disc like images in his short term memory, the higher dimensional shape appears in his perceptual space. In this way Mr. Square overcomes his dimensional impairment. Attractors in human perceptual space reach about 5D.(Authors' diagram.)

As the discs stack up, the Sphere takes on the appearance of a Honey Spoon. We believe that a mechanism similar to this allows our 3D brain to comprehend our own higher dimensional perceptual space. Our Honey Spoon theory closely resembles the cinematographic theory of consciousness.

It is easy to appreciate that during vertebrate evolution the dimensional capacity of our perceptual space increased in parallel with our ability to engage in higher thought. Professor Walter Freeman's group at Berkeley has shown that the cerebral cortex is able to organize itself and that it undergoes phase transitions which are analogous to boiling and condensing water. The transitions, which may occur about every 25 ms, are thought to represent frames of information which constitute the cinematographic frames of

Consciousness. Freeman attributes the 25 ms timing to the natural firing rates between excitatory and inhibitory pyramidal neurons in the cerebral cortex (8 and 9).

"Honey spoon" theory explaining the Conscious Moment.

Successive "frames" of AM Wave Packets are added every ~25ms. They fade after ~400ms.

The "Honey spoon" is a multidimensional construct (Up to 5D), within perceptual space ~400ms. "long", and being continually refreshed at the "front" while fading at the "back end".

The "Honey spoon" provides a "location" for Consciousness and a site for binding data. It frees the percipient from the constraint imposed by the event horizon of 3D physical space.

Figure 11. Honey Spoon Theory. (Artwork by author.)

The Honey Spoon Theory of consciousness thus provides a "location" for consciousness, a mechanism to explain how dimensionally deprived humans may access higher dimensions, and a simple explanation of "binding".

And now let us consider angels.

CHAPTER 2

Angels

They appear without warning and disappear without a trace. First, a distinction must be made between reality, illusion and hallucination. We have argued that the percept which Tatum enjoyed when she saw the elephant at the zoo was real to Tatum. The fact that the percept appeared in Tatum's non-physical perceptual space is quite beside the point. It is within Tatum's perceptual space that her own contact with reality resides. To avoid sensory overload the brain only samples part of the local environment and the resulting smorgasbord provides for Tatum's concept of reality. She does not need to "see" X-Rays and microwaves while looking at the elephant.

When Tatum returns home from the zoo she might see a crumpled blanket in a dark corner of her room. That represents physical reality. If her senses fool her and she mistakes the blanket for her dog, that is an illusion. If she sees a dog and there is nothing there, that is a hallucination. Those who do not believe in angels are quick to point out that reports of angelic encounters probably came from witnesses who were deluded or hallucinating, even supposing that the encounter was reported honestly. Now a single deluded prophet is not outside the bounds of possibility, especially if he has been fasting for 40 days and 40 nights: this may explain why artful angels often appeared to groups of witnesses simultaneously. A starving prophet suffering a delusion is one thing, but a well fed group of down to earth shepherds watching their flocks by night and all suffering a simultateous delusion quite another.

We will start with abbreviated definitions of angels:

Hebrew.

A messenger, one sent with a message, a prophet, a priest. A messenger from God acting as an interpreter and declaring what is right.

Anglo Catholic.

A spiritual being superior to humans in power and intelligence.

Roman Catholic.

An **angel** is a pure spirit created by God. The Old Testament theology included the belief in angels: the name applied to certain spiritual beings or intelligences of heavenly residence, employed by God as the ministers of His will.

Islam.

Angels are heavenly beings mentioned many times in the <u>Quran</u> and <u>hadith</u> literature. Unlike humans or <u>jinn</u>, they have no biological needs and therefore no lower desires predicated by animal nature; consequently, they may be described as creatures of pure reason, who though endowed with <u>free will</u> are not subject to temptation. (Wikipedia, the free encyclopedia.)

Thus, the most important Abrahamic religions would probably agree that an angel is a messenger from God. Angels are purely spiritual beings, created by God, and superior to humans in power and intelligence. They have no biological needs and are therefore creatures of pure reason.

Before embarking on a survey of a variety of angelic appearances which are recorded in the Bible, it would be amiss not to mention the greatest Angelic scholar ever, Saint Thomas Aquinas "The Dumb Ox".(c.1225-1274). Born into a noble Neapolitan family, Thomas chose the life of a mendicant friar. Lumbering and shy- his classmates dubbed him "the Dumb Ox"- he led a revolution in Christian thought. Posessed of the rarest brilliance, he found the highest truth in the humblest object (10).

His greatest work, Summa Theologiae or Summa Theologica (5) was written between 1265 and 1274. It consisted of three parts, the First Part describes God's existence and nature, the creation, angels and the nature of man. Of the 119 questions considered, questions 50-64 concern angels. The Second Part deals mainly with morality and virtues and the Third part deals with the person and work of Christ, the sacraments and the end of the world. His thorough description of angels earned him the name of "The Angelic Doctor".

James Collins wrote in a dissertation about St. Thomas Aquinas (11).

"What are angels and how do we know of their existence?

Thomas gives an argument that the perfection of the universe requires the existence of intellectual creatures. Since God intends the good for His creation, he intends that it be like Himself. And since an effect is most like its cause when it shares with it the feature whereby it was caused, God's creation must contain something with intellect and will since that is how God creates, i.e. by first knowing it and loving it into being.

Hence the perfection of the universe requires that there should be intellectual creatures. Now to understand cannot be the action of a body, nor of any corporeal power.... Hence the perfection of the universe requires the existence of an incorporeal creature. (ST Ia 50, 1)

However, since humans are intellectual creatures, as he indicates at the end of this very argument, the need for some intellectual creatures is not sufficient to give us knowledge of the existence of purely intellectual creatures which the angels are.

Since Sacred Scripture does speak definitively about the existence of angels, it belongs to Sacred Doctrine, i.e. theology, to treat of angels in a truly scientific manner. The divine science has the intellectual tools (faith in Scripture) to establish both the fact of angels and their nature (ST Ia, 1, 3). Having accepted on faith that angels exist, or taking their existence to be purely hypothetical, one can still draw certain philosophical conclusions about their nature. Thomas' words in the Summa are an excellent guide for how one can think clearly about the angelic hosts. For Thomas, given that angels are intellectual creatures, they must be pure spirit, i.e. self-subsistent forms. They are completely incorporeal; they are in no way material, and have no bodies of any kind.(Ia 50, 2) ".

What follows is a literal but abridged translation of Question 51. The reply has been shortened considerably due to constraints on space and this short extract is purely intended to give the reader a sample of St. Thomas' writing style

From Summa Theologiae.

"Question 51. The angels in comparison with bodies

Article 1. Whether the angels have bodies naturally united to them?

Objection 1. It would seem that <u>angels</u> have bodies <u>naturally</u> united to them. For <u>Origen</u> says (Peri Archon i): "It is <u>God's</u> attribute alone--that is, it belongs to the Father, the Son, and the <u>Holy Ghost</u>, as a property of <u>nature</u>, that He is understood to <u>exist</u> without any material <u>substance</u> and without any companionship of corporeal addition." <u>Bernard</u> likewise says (Hom. vi. super Cant.): "Let us assign incorporeity to <u>God</u> alone even as we do <u>immortality</u>, whose <u>nature</u> alone, neither for its own sake nor on account of anything else, needs the help of any corporeal organ. But it is clear that every <u>created</u> spirit needs corporeal <u>substance</u>." <u>Augustine</u> also says (Gen. ad lit. iii): "The <u>demons</u> are called animals of the atmosphere because their <u>nature</u> is akin to that of aerial bodies." But the <u>nature</u> of <u>demons</u> and *angels* is the same. *Therefore angels have bodies naturally united to them.*

Objection 2. Further, *Gregory* (Hom. x in Ev.) calls an *angel* a rational animal. But every animal is composed of body and *soul*. Therefore *angels* have bodies *naturally* united to them.

Objection 3. Further, life is more perfect in the *angels* than in *souls*. But the *soul* not only lives, but gives life to the body. Therefore the *angels* animate bodies which are *naturally* united to them.

On the contrary, Dionysius says (Div. Nom. iv) that "the *angels* are understood to be incorporeal."

I answer that, The *angels* have not bodies *naturally* united to them. For whatever belongs to any *nature* as an *accident* is not found universally in that *nature*; thus, for instance, to *have wings, because it is not of the <u>essence</u> of an animal, does not belong to every animal" (5).*

To summarize St. Thomas Aquinas; angels are pure spirit, self-subsistent forms. They are pure intellects completely incorporeal and not united to a body. They may occasionally "borrow" a body when they are to be seen commonly by all. (I do not think this is necessary, we will deal with this later). Their primary purpose is to offer a preview of the life to come.

We will now consider appearances of angels in the Bible (12).There are about 300 references to angelic appearances, 17 of them in the New Testament. This is not an exhaustive list intended to cover every appearance, rather it is an attempt to highlight well known stories and to emphasize the different ways that angels present themselves.

We have also included a few references to Jesus as they are unique. Only then will it be possible to see if all eventualities are covered by our theory.

The Old Testament:

The Fall of Adam and Eve.

{3:23} Therefore the LORD

God sent him forth from the garden of Eden, to till the ground from whence he was taken. {3:24} So he drove out the man; and he placed at the east of the garden of Eden Cherubim, and a flaming sword which turned every way, to keep the way of the tree of life.

Genesis 3:23-24

The Three Angels Appear to Abraham.

{18:1} And the LORD appeared unto him in the plains of Mamre: and he sat in the tent door in the heat of the day; {18:2} And he lift up his eyes and looked, and, lo, three men stood by him: and when he saw [them,] he ran to meet them from the tent door, and bowed himself toward the ground, {18:3} And said, My Lord, if now I have found favour in thy sight, pass not away, I pray thee, from thy servant:

Genesis 18:1-3

• • • • • • • • ● ● ● ● • • • •

Jacob's Ladder

{28:10} And Jacob went out from Beer-sheba, and went toward Haran. {28:11} And he lighted upon a certain place, and tarried there all night, because the sun was set; and he took of the stones of that place, and [put] them for his pillows, and lay down in that place to sleep. {28:12} And he dreamed, and behold a ladder set up on the earth, and the top of it reached to heaven: and behold the angels of God ascending and descending on it.

Genesis 28:10-12

· · · · · · · ● ● ● **●** ● ● · · · · · · ·

God Sends an Angel to Lead Moses

{23:20} Behold, I send an Angel before thee, to keep thee in the way, and to bring thee into the place which I have prepared. {23:21} Beware of him, and obey his voice, provoke him not; for he will not pardon your transgressions: for my name [is] in him. {23:22} But if thou shalt indeed obey his voice, and do all that I speak; then I will be an enemy unto thine enemies, and an adversary unto thine adversaries.

Exodus 23:20-22

Daniel in the lion's den.

{6:19} Then the king arose very early in the morning, and went in haste untothe den of lions. {6:20} And when he came to the den, he cried with a lamentable voice unto Daniel: [and] the king spake and said to Daniel, O Daniel, servant of the living God, is thy God, whom thou servest continually, able to deliver thee from the lions? {6:21} Then said Daniel unto the king, O king, live for ever. {6:22} My God hath sent his angel, and hath shut the lions mouths, that they have not hurt me: forasmuch as before him innocency was found in me; and also before thee, O king, have I done no hurt.

Daniel 6:19-22

• • • • • • • • ● • • • • • • • •

THE NEW TESTAMENT.

An Angel Appears to Joseph.

{1:18} Now the birth of Jesus Christ was on this wise: When as his mother Mary was espoused to Joseph, before they came together, she was found with child of the Holy Ghost. {1:19} Then Joseph her husband, being a just [man,] and not willing to make her a publick example, was minded to put her away privily. {1:20} But while he thought on these things, behold, the angel of the Lord appeared unto him in a dream, saying, Joseph, thou son of David, fear not to take unto thee Mary thy wife: for that which is conceived in her is of the Holy Ghost. {1:21} And she shall bring

forth a son, and thou shalt call his name JESUS: for he shall save his people from their sins.

Matthew 1:18-21

· · · · · · · ● ● ● ● ● ● · · · · · · ·

The Angel Gabriel appears to Mary.

{1:26} And in the sixth month the angel Gabriel was sent from God unto a city of Galilee, named Nazareth, {1:27} To a virgin espoused to a man whose name was Joseph, of the house of David; and the virgin's name [was] Mary. {1:28} And the angel came in unto her, and said, Hail, [thou that art] highly favoured, the Lord [is] with thee: blessed [art] thou among women.

Luke 1:26-28

• • • • • • • • ● ● • • • • • •

An Angel appears before the shepherds..

{2:8} And there were in the same country shepherds abiding in the field, keeping watch over their flock by night. {2:9} And, lo, the angel of the Lord came upon them, and the glory of the Lord shone round about them: and they were sore afraid. {2:10} And the angel said unto them, Fear not: for, behold, I bring you good tidings of great joy, which shall be to all people. {2:11} For unto you is born this day in the city of David a Saviour, which is Christ the Lord. {2:12} And this [shall be] a sign unto you; Ye shall find the babe wrapped in swaddling clothes, lying in a manger. {2:13} And suddenly there was with the angel a multitude of the heavenly host praising God,

and saying, {2:14} Glory to God in the highest, and on earth peace, good will toward men. {2:15} And it came to pass, as the angels were gone away from them into heaven, the shepherds said one to another, Let us now go even unto Bethlehem, and see this thing which is come to pass, which the Lord hath made known unto us. {2:16} And they came with haste, and found Mary, and Joseph, and the babe lying in a manger. {2:17} And when they had seen [it,] they made known abroad the saying which was told them concerning this child. {2:18} And all they that heard [it] wondered at those things which were told them by the shepherds. {2:19} But Mary kept all these things, and pondered [them] in her heart. {2:20} And

*the shepherds returned, glorifying
and praising God for all the things
that they had heard and seen, as it
was told unto them.*

Luke 2:8-20

• • • • • • • ● • • • • • • • •

The Resurrection of Jesus .

{20:11} But Mary stood without at the sepulchre weeping: and as she wept, she stooped down, [and looked] into the sepulchre, {20:12} And seeth two angels in white sitting, the one at the head, and the other at the feet, where the body of Jesus had lain. {20:13} And they say unto her, Woman, why weepest thou? She saith unto them, Because they have taken away my Lord, and I know not where they have laid him. {20:14} And when she had thus said, she turned herself back, and saw Jesus standing, and knew not that it was Jesus. {20:15} Jesus saith unto her, Woman, why weepest thou? whom seekest thou? She, supposing him tobe the gardener, saith unto him,

Sir, if thou have borne him hence,
tell me where thou hast laid him,
and I will take him away.

John 20:11-15

• • • • • • • • • ● ● • • • • • • • •

Jesus appears to his deciples.

{20:19} Then the same day at evening, being the first [day] of the week, when the doors were shut where the disciples were assembled for fear of the Jews, came Jesus and stood in the midst, and saith unto them, Peace [be] unto you. {20:20} And when he had so said, he shewed unto them [his] hands and his side. Then were the disciples glad, when they saw the Lord. {20:21} Then said Jesus to them again, Peace [be] unto you: as [my] Father hath sent me, even so send I you. {20:22} And when he had said this, he breathed on [them,] and saith unto them, Receive ye the Holy Ghost: {20:23} Whose soever sins ye remit, they are remitted unto them; [and]

whose soever [sins] ye retain, they are retained. {20:24} But Thomas, one of the twelve, called Didymus, was not with them when Jesus came. {20:25} The other disciples therefore said unto him, We have seen the Lord. But he said unto them, Except I shall see in his hands the print of the nails, and put my finger into the print of the nails, and thrust my hand into his side, I will not believe. {20:26} And after eight days again his disciples were within, and Thomas with them: [then] came Jesus, the doors being shut, and stood in the midst, and said, Peace [be] unto you. {20:27} Then saith he to Thomas, reach hither thy finger, and behold my hands; and reach hither thy hand, and thrust [it] into my side: and

*be not faithless, but believing.
{20:28} And Thomas answered
and said unto him, My Lord and
my God. {20:29} Jesus saith unto
him, Thomas, because thou hast
seen me, thou hast believed: blessed
[are]they that have not seen, and
[yet] have believed.*

John 20:19-29

· · · · · · · · · ● · · · · · · · · ·

Saul's Conversion.

{9:1} And Saul, yet breathing out threatenings and slaughter against the disciples of the Lord, went unto the high priest, {9:2} And desired of him letters to Damascus to the synagogues, that if he found any of this way, whether they were men or women, he might bring them bound unto Jerusalem. {9:3} And as he journeyed, he came near Damascus: and suddenly there shined round about him a light from heaven: {9:4} And he fell to the earth, and heard a voice saying unto him, Saul, Saul, why persecutest thou me? {9:5} And he said, Who art thou, Lord? And the Lord said, I am Jesus whom thou persecutest: [it is] hard for thee to kick against the pricks. {9:6} And

he trembling and astonished said, Lord, what wilt thou have me to do? And the Lord [said] unto him, Arise, and go into the city, and it shall be told thee what thou must do. {9:7} And the men which journeyed with him stood speechless, hearing a voice, but seeing no man. {9:8} And Saul arose from the earth; and when his eyes were opened, he saw no man: but they led him by the hand, and brought [him] into Damascus. {9:9} And he was three days without sight, and neither did eat nor drink. {9:10} And there was a certain disciple at Damascus, named Ananias; and to him said the Lord in a vision, Ananias. And he said, Behold, I [am here,] Lord. {9:11} And the Lord [said] unto him, Arise, and go into the street which is called Straight, and enquire in the house of Judas for [one] called Saul, of Tarsus: for, behold, he prayeth, {9:12} And hath seen in a vision a man named Ananias coming in, and putting [his] hand on him, that he might receive his sight. {9:13} Then Ananias answered, Lord, I have heard by many of this man, how much evil he hath done to thy saints at Jerusalem: {9:14} And here he hath authority from the chief priests to bind all that call on thy name. {9:15} But the Lord said unto him, Go thy way: for he is a chosen vessel unto me, to bear my name before the Gentiles, and kings, and the children of Israel: {9:16} For I will shew him how great things he must suffer for my name's sake. {9:17} And Ananias went his way, and entered into the house; and putting his hands on him said, Brother Saul, the Lord, [even] Jesus, that appeared unto thee in the way as thou camest, hath sent me, that thou mightest receive thy sight, and be filled with the Holy Ghost. {9:18} And immediately there fell from his eyes as it had been scales: and he received sight forthwith, and arose, and was baptized. {9:19} And when he had received meat, he was strengthened. Then was Saul certain days with the disciples which were at Damascus. {9:20} And straightway he preached Christ in the synagogues, that he is the Son of God. {9:21} But all that heard [him] were amazed, and said; Is not this he that destroyed them which called on this name in Jerusalem, and came hither for that intent, that he might bring them bound unto the chief priests? {9:22} But Saul increased the more in strength, and confounded the Jews which dwelt at Damascus proving that this is very Christ.

Acts 9:1-22

● ● ● ● ● ● ● ● ● ● ● ● ● ● ● ● ● ●

The Angel Rescues Peter from Jail.

{12:6} And when Herod would have brought him forth, the same night Peter was sleeping between two soldiers, bound with two chains: and the keepers before the door kept the prison. {12:7} And, behold, the angel of the Lord came upon [him,] and a light shined in the prison: and he smote Peter on the side, and raised him up, saying, Arise up quickly. And his chains fell off from [his] hands. {12:8} And the angel said unto him, Gird thyself, and bind on thy sandals. And so he did. And he saith unto him, Cast thy garment about thee, and follow me. {12:9} And he went out, and followed him; and wist not that it was true which was done by the angel; but thought he

saw a vision. {12:10} When they were past the first and the second ward, they came unto the iron gate that leadeth unto the city; which opened to them of his own accord: and they went out, and passed on through one street; and forthwith the angel departed from him.

Acts of the Apostles 12:6-10

• • • • • • • ● ● ● ● ● ● ● • • • •

Letter of St. Paul to the Hebrews.

{13:1} Let brotherly love continue.
{13:2} Be not forgetful to entertain
strangers: for thereby some have
entertained angels unawares.

Hebrews 13:1-2

• • • • • • • • ● ● • • • • • •

These accounts of angelic appearances are just a few examples which are well known, and serve to illustrate the different conditions under which angels operate. A comprehensive account of their *modus operandi* should offer a reasonable explanation for all these conditions.

In Genesis 18: 1-3, Abraham sees three angels appear outside his tent, but in Genesis 28: 10-12, Jacob sees angels in a dream. In Exodus 23: 20-22, God introduces an angel to Moses and in Daniel 6: 19-22, an angel protects Daniel from the lions.

In the New Testament, Matthew 1: 18-21, an angel appears to Joseph in a dream but in Luke 2: 8-20, an angel appears to multiple shepherds. In John 20: 11-15, Mary sees two angels in the tomb of Jesus. The conversion of Saul, Acts 9: 1-22, is curious because while Saul sees a blinding light and hears a voice, his companions only hear the voice. In Acts 12: 6-10, Peter's release from prison illustrates an angel in action while in Hebrews 13: 1-2 Paul warns us that angels may not always be recognizable.

It is now time to consider if these accounts from scripture may be reconciled with modern science, especially up to date theories of how human consciousness works.

CHAPTER 3

Modus Operandi
of angels

We have argued that angels are messengers from God and that they are incorporeal; that is to say that they have no physical bodies. They are powerful, highly intelligent and seem to appear and disappear as the situation warrants. They appear at important times in human history when God's creatures are confused, bewildered or in trouble. The Ten Commandments should cover most eventualities but it would seem that occasionally man needs a little more direct encouragement. They appear in dreams and in broad daylight, singly or in groups, and appear unannounced. They know that their sudden appearance causes fear and terror and they often reassure the witness to "Fear Not"; which is hardly the signature of a hallucination. One of the ways their presence may be understood is to hark back to our modern theories of consciousness.

I believe that angels exist in a higher dimension than the three with which we are all accustomed.

Now we come to our astonishing hypothesis which arises from a fusion of historial reports and modern theories of consciousness. When I started writing this book I had no clear idea how it would end. It seemed reasonable to combine ancient texts with modern theories of consciousness and see how an appreciation of the higher dimensions found in the neurodynamics of perceptual space correlated with ancient recordings of angelic appearances. Our hypothesis hinges on two assuptions, first, that angels reside in and visit us from a higher dimension. Secondly, as we have demonstrated, that perceptual space is not only non-physical but may exist in up to 5D. Evolution has "hijacked" higher dimensional perceptual space precicely to accommodate 3D vision, hearing, smell and taste etc. simultaneously. It is only in this way that we may respond rapidly to dangers and opportunities in our environment. Our hypothesis states that angelic appearances may arise either in physical space or directly in the perceptual space of the witness.

It is time to return to Flatland. Mr. Square is unchanged. He still lives in a polygon. (Fig 12.) Access is via an opening which appears when one side swings out (Fig 13.)

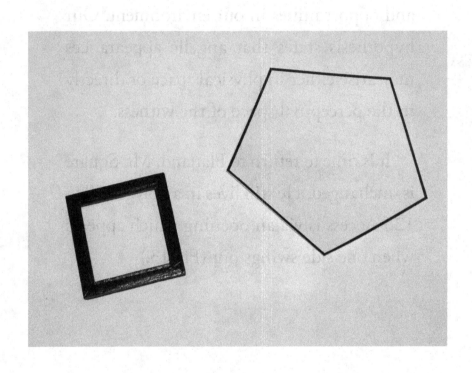

Figure 12. Mr. Square owns a flat pentagonal house which he keeps locked. (Authors' photo).

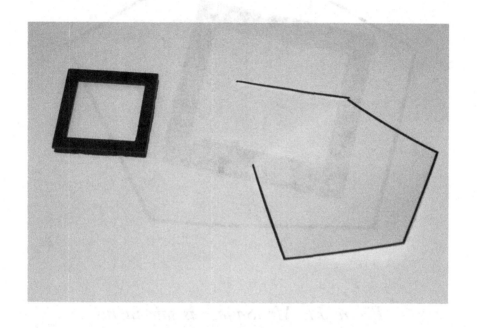

Figure 13. Mr. Square obtains access as one side swings open. (Authors' photo).

Mr. Square is now safely indoors. (Fig 14.)

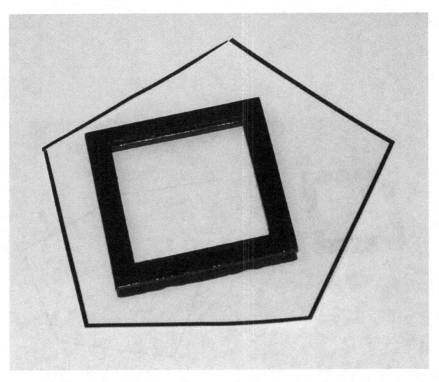

Figure 14. Mr. Square is safe in his house. Remember that if Sphere or an angel comes to vist they may drop in from above. This would seem to Mr.Square that they appeared from nowhere. (Authors' photo).

Now suppose that an angel ("A") appears to deliver a message. (Fig.14).

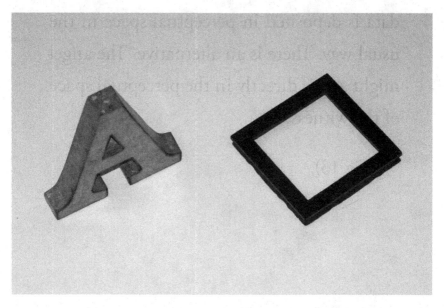

Figure 12. An angel makes a visit.
(Authors' photo).

The angel is a higher dimensional being than the square who sees nothing until the angel touches Flatland. As the angel touches Flatland, Mr.Square suddenly sees a bright light appearing from nowhere. In this scenario, Mr. Square sees the angel using his normal sensory apparatus. From here, the data is deposited in perceptual space in the usual way. There is an alternative. The angel might arrive directly in the perceptual space of the witness.

(Fig 13).

Figure 13. The non-physical angel may appear directly in the perceptual space of the witness. (Authors' photo).

What happens when an angel appears to more than one person? (Figs 14 and 15).

Figure 14. An angel appears to more than one witness. They may use their normal senses to perceive the visitor. (Authors' photo).

Figure 15. The visiting angel appears in the perceptual spaces of more than one witness. A multidimensional being may morph into any shape which suits the job. (Authors' photo).

Let us consider some of our examples of angelic visits and try to categorize them.

A. *One on one visitation.* In Luke 1: 26-28, the angel Gabriel visited Mary to announce the birth of Jesus which was to occur about 3 months later. In this example Gabriel could have appeared in Mary's perceptual space or as a vision in physical space which Mary would process with her eyes and hearing. From Mary's point of view, she might not be able to tell the difference unless she closed her eyes while Gabriel was talking. Then she would hear but not see for a moment. From Gabriel's vantage point, it would probably be more logical to appear directly in Mary's perceptual space. However, angelic logic is not the subject of this work.

B. *Appearing in a dream.* In Genesis 28: 10-12 we read the account of Jacob's Ladder. The angels appeared in a dream and we would assume that the angels appeared in Jacob's perceptual space.

C. *Appearing before a crowd.* In Luke 2: 8-20 an angel appears before the shepherds. Why shepherds and why a group and not one shepherd? First, shepherds were far removed from those who might have aspirations to the throne and they certainly had no political axes of their own to grind. This provided them with the mantle of impartial and credible witnesses. The appearance before a crowd reduces the risk of a single witness being accused of being crazy or drunk. The whole group of them adds to the celebration of Jesus' birth. The angel might still have appeared in each individual shepherd's perceptual space (Fig. 15) simultaneously by virtue of his higher dimensional form, but in this case the appearances could just as easily have been "out there" to be enjoyed by the shepherds' own senses.

D. *Sauls Conversion.* In Acts 9:1-22 we read of Saul's conversion. Although Jesus is talking to Saul I include this account because it presents features not seen elsewhere. While Saul sees a dazzling light and hears a voice, his companions only hear the voice. I think it probable that the light appeared in Saul's perceptual space while the sound appeared in physical space so that the witnesses would know that something truly remarkable was happening and that Saul was not having a stroke. This example is good evidence that angelic appearences may result in an appearance in either physical space or perceptual space depending on the ciecumstanses.

E. *Jesus appears in a locked room.* In John 20: 19-29 Jesus appears to his diciples on two occasions a week apart. Students of Flatland will have no difficulty in understanding how a higher dimensional being may suddenly appear inside a locked room. He may have appeared in physical space or in the diciples' perceptual spaces. However, in this account of a post resurrection appearance he tells Thomas to feel his wounds. Also in John 21: 9-14 when Jesus becons to his diciples to join him in sharing grilled fish for breakfast, he wants to emphsise that he is really resurrected and that they are not seeing a ghost. In these examples it would seem more logical for Jesus to appear in physical space.

F. *An angel rescues Peter from prison.* In Acts12:6-10 the angel strikes Peter to wake him up, Peter's chains fall off and the prison door opens by itself. This is not just a messenger but *"action angel"*. Because of the physical interaction with the prison, it would seem that an appearance in physical space would be more appropriate than the angel merely appearing in Peter's perceptual space.

It would appear from these different examples that angels may appear in dreams, as a physical presence in a dusty jail cell, or almost anything in between.

The cinematographic Honey Spoon theory of consciousness, in conjunction with examples from Flatland, form a stable platform from which to study angelic appearances. The concept of a separation between physical space and perceptual space brings me to my one point of disagreement with St. Thomas Aquinas. In our category "C", when an angel appeared before a crowd, St. Thomas suggests that angels might "borrow a body when they are to be seen commonly by all". If, as we suggest, angels exist in a higher dimension and only become apparent to us when they intersect our 3D physical space or alight in our perceptual space, there is no need for them to "borrow" a body at all because as we showed in Figure 15, a higher dimensional form may easily spread itself out to interact with multiple witnesses simultaneously.

The discriminating reader will deduce that I have not proved anything at all. I agree. The purpose of this small volume is to suggest that angelic visits were and are feasible. They should not, therefore, be relegated to the world of fiction. I will allow St. Paul the last word;

Let brotherly love continue. Be not forgetful to entertain strangers: for thereby some have *entertained angels unawares.* ***Hebrews 13:1-2***

References

1. Walling PT, Hicks KN: Brain, Mind, Cosmos: The Nature of Our Existence and the Universe. Sages and Scientists Series, Part 1.Editor Deepak Chopra: Chapter 30: The Dynamics of Consciousness.2013.

2. Walling PT: The Neurodynamics of Consciousness, Chapter 7 in Consciousness. Integrating Eastern

and Western Perspectives. Editors Prem Saran Satsangi and Stuart Hameroff. Pp 147-166. New Age Books, New Delhi (India) 2016.

3. Walling PT, Hicks KN; Nonlinear changes in brain dynamics during emergence from Sevoflurane anesthesia. ANESTHESIOLOGY 2006; V 105; pp927-35.

4. Walling PT, Hicks KN: Consciousness: Anatomy of the Soul, Authorhouse. 2009.

5. St. Thomas Aquinas: Summa Theologica, 1920. Fathers of the English Dominican Province.

6. Russell B: An Outline of Philosophy, London, Routledge, 1996, p109.

7. Abbott EA: Flatland. A Romance in many Dimensions. New York. Cosimo Classics. 2007. (Original printing 1884).

8. Freeman WJ, Rogers LJ: Fine resolution of Analytic Phase reveals Episodic Synchronization by State Transitions in Gamma EEG's, J Neurophysiol, 2002, 87, pp 937-945.

9. Freeman WJ, Burke BC, Holmes MD: Berkeley Lecture, 2003, Human Brain Mapping.http://lecture,Berkeley.edu/ wjf/EE Scalp Hilbert Phase.pdf

10. Chesterton GK: Saint Thomas Aquinas "The Dumb Ox" New York. Image Books. 1933(1956) Back cover.

11. Collins J: The Thomistic Philosophy of the Angels, a dissertation,

(Washington, DC: Catholic University Press, 1947).

12. (http://www.ulc.org/wp-content/ uploads/2012/09/King-James-Bible-KJV-Bible-PDF.pdf)

Printed in the United States
by Bookmasters

Printed in the United States
By Bookmasters